Table of Contents

Introduction

Welcome to **MathsBlasters – Algebra** and well done for having made a good choice.

This is part of a series of books that have been designed to help people overcome the difficulties they are experiencing with maths; especially achieving the grade they are trying to achieve at foundation level GCSE.

Also in the series are:

MathsBlasters – Number
MathsBlasters – Geometry
MathsBlasters – Statistics
MathsBlasters – Success Strategies for Maths
GCSE MathsBlaster (the complete exam book)

You are probably looking at the MathsBlaster series of books for one reason:

You have been struggling with GCSE maths and either you have been getting mock scores which are below the grade you want, or you are doing a re-sit course and you are concerned that you aren't making enough progress.

And the reason you have chosen this specific book is that you are having difficulty with the various 'algebra' skills involved in maths, particularly at GCSE level and you would like to improve.

If you follow the advice that I give, throughout the book, you will soon be on the way to making the improvements, in your algebra skills, that you really want to make.

Algebra is a vital part of maths, which is why I have written this book; to help people like you overcome the obstacles they face in trying to improve their maths performance.

Apart from the algebra skills, strategies and techniques that this book covers, you will sometimes see reference to **up to the staples'** which is a strategy to help you get as many marks as you can, by concentrating on the basics.

I go into **up to the staples'** in more detail through my full course book, **GCSE MathsBlaster** but the concept is really quite simple:

Look at an exam paper and particularly a non-calculator paper and you will see that the questions in the first half of the paper (where the staples are; right in the middle) are generally really easy and often make up around half the marks on that paper.

In most cases, the questions **up to the staples'** involve being able to use all the basic algebra skills and especially being able to solve simple equations, simplify expressions and answer questions involving brackets.

So whenever you see **up to the staples'** in one of my books, that is what it is referring to.

It's all about being able to do the basics well and having a strategy to check if your answers make sense.

Another thing you will see in this book quite often is me reminding you to do your **ABC** which means:

Always Be Checking

In other words – NEVER take your answer for granted. NEVER assume that your answer is correct and then move on to the next question, without checking.

You must always remember your ABC

The Basic Facts About Algebra

Here it is; everyone's favourite topic.

Algebra!!

When children move up from primary school to secondary school, people so often tell them that when they get to secondary school they are going to have to do algebra and . . . .

It's really, really hard.

Whenever I talk to young people about maths and which bit of maths they feel they struggle with most, they will invariably tell me that they struggle with algebra.

Which is a shame because algebra is dead easy. In fact, it is one of the easiest maths topics.

So, if I am going to help you to become good enough to achieve the score you need, on the exam paper, then I am going to have to get you thinking that **algebra is easy**!

Let's get started then.

I am going to show you some algebra questions that appeared on a recent GCSE foundation maths paper and 'walk you through them'

And by the way; all the work you do should be in a revision pad, where you should make detailed notes and use **colour** plus **BOLD, HIGHLIGHTED examples**

Doing that will help you to remember and recall key information more easily.

See my website: **www.mathsblasters.co.uk** for more ideas

Let's begin by reminding you of one of the real basics of algebra:

When a number and a letter are written next to each other it means they are being multiplied.

For example: 4a simply means 4 x a or 4 lots of a.

You need to keep that simple fact in your mind so that if you see a number and a letter together, you will know they are a multiplication.

5b means 5 x b

6p means 6 x p

11d means 11 x d

The next important rule for algebra is what the word **simplify** means.

Simplify means make it simpler or write something more simply. Another way to say this in algebra would be 'make it shorter'

You also need to remember to use your multiplication tables, as solving algebra problems is really easy once you know your multiplication facts.

Until you have mastered your tables, use a multiplication grid to help you.

I have put a copy of a completed 12 x 12 table on the next page.

But, you DO need to learn these tables for yourself.

If you visit my website: www.mathsblasters.co.uk you will find more information and help, with your tables.

X	1	2	3	4	5	6	7	8	9	10	11	12
1	1	2	3	4	5	6	7	8	9	10	11	12
2	2	4	6	8	10	12	14	16	18	20	22	24
3	3	6	9	12	15	18	21	24	27	30	33	36
4	4	8	12	16	20	24	28	32	36	40	44	48
5	5	10	15	20	25	30	35	40	45	50	55	60
6	6	12	18	24	30	36	42	48	54	60	66	72
7	7	14	21	28	35	42	49	56	63	70	77	84
8	8	16	24	32	40	48	56	64	72	80	88	96
9	9	18	27	36	45	54	63	72	81	90	99	108
10	10	20	30	40	50	60	70	80	90	100	110	120
11	11	22	33	44	55	66	77	88	99	110	121	132
12	12	24	36	48	60	72	84	96	108	120	132	144

Ok.

Let's do some algebra.

Let's start with the exam question I mentioned earlier:

3c = 18

I am going to start by rewriting the question.

You now know that 3c means 3 x c, so really the question is:

3 x c = 18

The *'correct'* algebra way to answer this is to say that if we divide 18 by 3 we will find out what c is.

So, $18 \div 3 = 6$ which means that c = 6

We can check this using our **ABC** of maths.

We have to **A**lways **B**e **C**hecking

Right.

So, we think the answer to 3c = 18 is that c = 6

That means 3 x 6 = 18

Yes; that's correct. My tables tell me that 3 x 6 = 18 so I ***know*** that c = 6

Another way of looking at this question is to go straight to your multiplication tables at the start.

3c = 18, which means that 3 x c = 18 so I just check what I know about my three times table.

From my tables, I know that 6 x 3 = 18 so that tells me straight away that the answer is 6.

So, let's do one of our own.

4a = 20 (we **know** that this means 4 x a = 20)

Use your 4 times table to work out the answer.

What do we times 4 by to get 20?

Got it?

Well done! The answer is indeed 5

4 x 5 = 20 which means that a = 5

Try this one, then look at the next page to see if you got it right:

5a = 30

My 5 times table tells me that 6 x 5 = 30, so that tells me the answer to 5a = 30

The answer is 6

a = 6

That's all there is to it for the basics of our first bit of algebra.

You can now solve an equation. No problem. Right?

So, let's try one that has a little bit more in the question.

Again, here is an exam question:

7d + 16 = 51

Let's start off by breaking this question down into plain English.

Something plus 16 = 51

What we could do here is to take the 16 away and see what we are left with.

If I take 16 away from 51 (51 – 16) the answer is 35

Now I know that the *something (the 7d bit) is 35 because 35 + 16 = 51*

So, if 7d = 35, what is the value of d?

Back to our tables.

In the 7 times table what do we multiply 7 by to get 35?

Correct!

5

So if 7 x 5 = 35 then I now *know* that d = 5

What do you think?

Easy?

Let's do another one, and then turn the page to see the answer

4a + 7 = 19

Let's start by writing it in English again.

Something plus 7 = 19

Take 7 away from 19 which leaves 12

So the *something* is 12

That means 4a = 12

My 4 times table tells me that 3 x 4 = 12

So the answer is 3

a = 3

As we have to Always Be Checking . . . . .

Let's see if 3 really is the correct answer.

4a + 7 = 19

So 4 x a + 7 = 19

If a = 3 then we would write 4 x 3 + 7 = 19

Well, 4 x 3 = 12

And 12 + 7 = 19

So the answer is spot on; **a = 3**

So far, what I have done in this section is to:

- **Use what we know from the times tables**

- **Use plain English to read the question**

- **Break the question into small, easy to understand bits**

That's all there is to it.

Solving something like 6a = 18 (6 times *something* is 18) is not difficult as long as you use those three bullet points.

That's dealt with the basic 'solving' bit of algebra that you need to be confident about if you are going to succeed at the exam paper.

Now for simplifying!

Simplifying is when we do exactly what it says on the tin.

We simplify

We make something simpler or, as I wrote earlier, we shorten it.

Imagine you were going through your wardrobe making a list of the different coloured T-shirts that you have.

In your wardrobe there are:

White T-shirts (W)

Blue T-shirts (B)

Red T-shirts (R)

Yellow T-shirts (Y)

You have 20 T-shirts and you make a note of the colour as you check them off:

3W, B, R, 2Y, 2R, 4W, 3B, 2R, 2Y

You can picture the wardrobe rail, can't you?

3 White then a blue then a red then 2 yellow then 2 red followed by 4 white and 3 blue plus 2 red and then 2 yellow T-shirts.

When I made the original list I used *algebra*.

Algebra is simply a maths way of writing things in a short way and that's what I did with the list of T-shirts. So, now let's complete the algebra.

The T-shirts we have are: 3W + B + R + 2Y + 4W + 3B + 2R + 2Y

There; that's what the list of T-shirts looks like if we use an **expression** to describe them.

An **expression** is just a maths way of writing a sentence.

In English we would say "I have 3 white T-shirts and a blue T shirt and a red T-shirt . . . . .. " and so on.

Well, in maths we just write all of that in a very simple way using letters and numbers.

An **expression** is basically like a maths sentence or phrase.

So, here's that list of T-shirts again:

3W + B + R + 2Y + 4W + 3B + 2R + 2Y

Now, what I want to do next, is *simplify* my T-shirt expression.

In your wardrobe, if you rearranged your T-shirts, you would put all the same colour T-shirts together. Next to each other.

That works in maths, too. So, let's do that.

3W + 4W + B + 3B + R + 2R + 2Y + 2Y

There. My expression now looks a bit like your wardrobe. T-shirts of the same colour are now next to each other.

So, all that is left for me to now, is *simplify.*

To *simplify* I just **add together things that are exactly the same**, which means I have:

7W + 4B + 3R + 4Y

That's it.

My expression was 3W + 4W + B + 3B + R + 2R + 2Y + 2Y

And I have simplified it to 7W + 4B + 3R + 4Y

Ready for an exam question now?

In the 2015 exam this was a two mark question!

For being able to simplify this expression you would be awarded two very, very easy marks.

Here goes . . ..

a) Simplify 7j + 6k – 5j + 4k   (2 marks)

Now then; how easy is **THAT?**

So, how many j are there? (7j -5j) = 2j

And how many k are there? (6k + 4k) = 10k

So, put those two answers together and you have 2j + 10k

And that's it.

Done.

Finished

Two really easy marks earned.

Let's do another one:

a) Simplify 6b + 3g – 2b + 4g

Same thing again; how many b are there? (6b – 2b) = 4b

And how many g are there? (3g + 4g) = 7g

So our answer is 4b + 7g

You have a go at this one. The answer is on the next
page

Simplify 7t + 2v – 3t + 6v

Let's check that one from the previous page

Simplify 7t + 2v − 3t + 6v

I can see **4t** (7t − 3t) and **8v** (2v + 6v)

So my answer is 4t + 8v   *Piece of cake!*

Let's move on to something **slightly** more challenging.

You will usually have to **work out the value** of something and the *something* might look like this:

Work out the value of 3a + 2b when a = 5 and b = 4

Yet again, this is one of those things that *looks* tricky but isn't. NOT if you tackle it properly.

So, first things first:

What does 3a mean? Well, 3a means 3 x a

What does 2b mean? Well, 2b means 2 x b

That's step 1. I have checked what the question actually means and I have also rewritten it, for a reason.

The reason is I want to do the question in two stages and then I will have the final answer. So, stage 1

If 3a means 3 x a then, if a = 5, I am really doing 3 x 5 = 15

If 2b means 2 x b then, if b = 4, I am really doing 2 x 4 = 8

Now all I have to do is put those two answers together 15 + 8 = 23, so the answer to **Work out the value of 3a + 2b when a = 5 and b = 4 is 23**

Your turn: **Work out the value of 5a + 4b when a = 6 and b = 3**

Once you have done that, turn to the next page.

So, let's start by rewriting the expression.

5a = 5 x a which is **5 x 6 = 30**

4b = 4 x b which is **4 x 3 = 12**

Now we add those two answers together.

30 + 12 = 42 so that's the answer to the question:

Work out the value of 5a + 4b when a = 6 and b = 3

There you are.

Now you can:

- Solve

- Simplify

- And work out the value of

And yes, I **know** we have been doing algebra but have you noticed something?

All we have been doing is multiplying, dividing, adding and subtracting.

You see, most of the maths we need, in order to do well in the exam paper, is all about the basics of number.

If you are going to succeed, you need to be able to add, subtract, multiply and divide; confidently and accurately.

So, by this point in the algebra chapter I hope you are thinking to yourself *"These algebra questions are a bit easy!"* because, that's what they are; **easy**!

Nothing to it; dead simple.

So, let's try something not so simple.

It's time to do a bit of multiplying out, which is sometimes called 'expanding'.

That's a key bit of language, right there.

Sometimes a question will tell you to **multiply out** and other times you will be asked to **expand**.

Both words mean the same thing and they appear with a question like this:

Multiply out 3(2x + 4) (May 2015 foundation paper 1)

The way this question works is that we have to multiply everything inside the brackets by the number on the outside.

Here, I have rewritten it on the next page using colour this time and made it larger:

3(2x + 4)

So, if we are going to **multiply everything inside the brackets by 3**, we will end up with:

(3 x 2x) + (3 x 4)

3 x 2x = 6x

3 x 4 = 12

So the answer is 6x + 12   That's it.

We cannot combine the two parts of the answer because they are not the same. There is no letter x attached to 12, so the answer is just:

6x + 12

Want to try one?

Simplify 4(3a + 5)

First of all let's rewrite the question:

4(3a + 5) becomes

(4 x 3a) + (4 x 5)

4 x 3a = 12a

4 x 5 = 20

Then put the two answers together to get 12a + 20

So, the answer to *"Simplify 4(3a + 5) is 12a + 20*

Now, let me tell you one of the most common errors I see when I mark questions like these.

Quite often pupils remember to multiply the first thing in the brackets but they forget to multiply the second thing, so I have often seen pupils write 12a + 5 as their answer to that last question.

They remembered to multiply the 3a part by 4 but they forgot to multiply the 5 by 4, which is why they end up with 12a + 5.

This is why you MUST remember the ABC of maths.

If you struggle with multiplying out brackets, even after my explanations, perhaps you would like to look at the grid method.

Let's see how we could use a grid to multiply out 6(3b + 5)

Draw a grid and put the algebra into it:

	3a	+5
6		

Can you see what I have done? I have put what is **inside** the brackets at the top and the bit ***outside*** the brackets at the side.

Now we can multiply everything together, just like we would if we were doing a routine multiplication

	3a	+5
6	18a	+30

All I did was multiply each of the bits at the top by 6 and the result   was 18a + 30

Let's do another:

This time we will try $4(5a^2 - 3)$

Start off with a grid and put the bit inside the brackets at the top and then the outside bit (4) at the side:

	$5a^2$	-3
4	$20a^2$	-12

My final answer; the bit inside the grid, is: $20a^2$ - 12

So, if you have previously found multiplying out brackets to be a bit tricky, maybe this different method, using a grid, will make it a bit easier for you.

When you do a 'multiply out' or 'expand' question, however, you MUST go back and check. You MUST ask yourself:

"Have I multiplied BOTH things inside the bracket?"

That is the **ABC** of maths that you have to use with a question like this and it is also a good reason why I have shown you how to break the question into two smaller questions and then, once you have worked out the answer to each little part, put those two smaller answers together to get the main answer.

So far, so good.

Hopefully this is going really well for you and you are thinking *"Hey! Algebra's not so tough after all."*

3, 7, 11, 15, 19 . . . . .

You've guessed it; we're moving on to sequences.

23?

If that's what you worked out as the next number in that sequence above, well done.

Mind you, it wasn't really that difficult, was it?

Well, sequences aren't *that difficult*

All you have to do, at the basic level, is:

1. Work out what the next one or two terms are

2. Explain why

So for the sequence above I would write 23 as the next term and then **add 4 or + 4** as my explanation.

For doing that I would get **two marks**!

And yet . . . ..

You would be amazed at the number of times I have marked a paper with a sequence like that, where the question asks for the next two terms and an explanation, only to find that the answer of some **very able** pupils looks a bit like this:

3, 7, 11, 15, 19, 23, 26

The rule is add 4 each time

Can you see the bit that amazes me?

19 + 4 = 23 (correct)

23 + 4 = 27 (NOT 26)

ABC

Making an error like that is simply down to carelessness and a failure to check the answer.

Goodness me, my grade C students certainly know how to add 4!

But a point I have made earlier is that sometimes, the answer is *so simple, so easy* that we just rush through and make a daft mistake.

And what's more, because it is such an easy question we don't give it the respect it deserves by going back to check it.

 Let me tell you something.

As a maths teacher I design and plan loads of resources and yet, whenever I prepare a resource for a class, *I never fail to check even the simplest answers.*

So if I am not so arrogant as to believe I don't need to check even the simplest answers, why should you, as a pupil, believe that **you** don't need to check?

I never take my answers for granted so neither should you!

For that sequence I would have double-checked. I would have gone back over the numbers to make sure; "does 19 + 4 = 23?" Yes. And then I would have asked myself "does 23 + 4 = 26?" at which point I would have thought "Don't be daft! 23 + 4 = 27"

And then I would have corrected the answer.

If a really good maths teacher always does his **ABC** then I think **you** should, too, don't you?

I am not going to dwell on that any further. I think the point has been made and I am not going to give you any more sequences until the end of the chapter.

What I AM going to do, though, is briefly touch on **nth term.**

An exam question may look like this:

"The expression for the nth term of a sequence is 4n – 2. Write down the first two terms of this sequence."

Now we get to the point where you might be starting to feel a little unsure.

That's understandable.

In my classroom, year after year, I find that **nth term** and the whole business of *"What is n?"* to be one that crops up all the time.

It's time to see if we can tackle this beast and tame it!

In this final algebra section (sequences) I am going to show you what **n** and **nth term** look like.

I am going to draw them.

First of all, some language.

In maths, the term number, **which we call *n*, is simply the position of a number in a sequence.**

So, in this sequence: 5, 10, 15, 20, 25

The first number in the sequence is **5**

The 4th number in the sequence is **20**

In maths, though, we would say 'the first *term* in the sequence is 5 and the 4th *term* in the sequence is 20.

Are you ok so far?

Let's check. In this sequence I would like you to decide which is the 3rd term in the sequence and which is the 5th term

2, 9, 16, 23, 30, 37, 44 : by the way, after I wrote that sequence, I went back over it and checked that each number really WAS 7 more than the previous one. I am definitely not cocky enough to publish a book without doing my ABC!

So, back to the sequence.

I assume you have worked out that the 3rd number in the sequence (the 3rd *term*) is 16 and that the 5th number in the sequence (the 5th *term*) is 30?

Time for another one

Here is a sequence

4, 10, 16, 22, 28, 34

This time, though, I am going to do something slightly different.

I am going to do something, which makes it really easy to tell each term in the sequence; **put the sequence into a table**.

Term	1	2	3	4	5	6
Sequence	4	10	16	22	28	34

Some people take to **nth term** like a duck to water and some don't.

If you are one of the ducks; fine!

On the other hand, if you struggle a bit with the whole *nth term* idea, this table method should be right up your street.

Now I am going to repeat the table but with one subtle difference

Term (n)	1	2	3	4	5	6
Sequence	4	10	16	22	28	34

Can you *Spot The Difference?*

Do you see that one, tiny character I have added in the second table?

Where it says 'Term' I have added (n) after it. Now you can easily see the term numbers. It is now really easy to see which is the second or fourth or sixth term in that sequence.

if you are at all unsure about nth term, I strongly suggest you use a table like that, for a while at least. Then, once you are more confident, you can do without it.

Let's give it a try with an exam question.

"The expression for the nth term of a sequence is 6n – 4. Write down the first three terms of the sequence."

If you have been struggling a bit with exam questions, we have probably reached the sort of question that you always hope to **NOT** see on a paper.

This is the type of question that a lot of pupils who are struggling with maths, hate.

Let's put that right in the next few minutes.

First of all I want you to get your revision pad (or any sheet of paper if you haven't got a revision pad yet) as we are going to draw a blank table to help.

Whenever you draw a table for showing a sequence I want you always write the rule right above the table, so you are clear about the job this table is going to do.

Now, let's go back to a bit earlier when I reminded you about what happens, in algebra, when a number and letter are written next to each other.

They are being multiplied.

Here's the table we are going to use, which I want you to copy before you read any further.

6n - 4

Term (n)	1	2	3	4	5	6
Sequence						

numbers to put in the empty boxes.

Remember; **6n** means **6 x n** so 6n-4 means **6 x n then take away 4.**

If **6n – 4 means 6 x n then take away 4, let's do that with the first *n*.**

The first n value in the table is 1

So, 6 x n – 4 means we do 6 x 1 then take away 4

6 x 1 = 6 then take away 4 becomes 6 – 4 = 2

Here's the table again with the first *n* filled in:

6n - 4

Term (n)	1	2	3	4	5	6
Sequence	2					

Let's do the second *n* now

The second n is 2 so now we have to do 6 x 2 then take away 4

6 x 2 = 12 and then take away 4 becomes 12 – 4 = 8

Let's add that to the table as the value for the 2nd *n*.

6n – 4

Term (**n**)	1	2	3	4	5	6
Sequence	2	8				

The exam question asked for the first three terms.

We have done 2 so now let's do the third and finish the question

I bet you know what I am going to do next?

The third term in the table, the third *n* is 3 so let's do the calculation:

6 x n – 4 this time means 6 x 3 then take away 4

6 x 3 = 18 and then take away 4 = 18 – 4 = 14

6n - 4

Term (**n**)	1	2	3	4	5	6
Sequence	2	8	14			

Do you want to complete terms 4, 5 and 6?

6n – 4

Term (**n**)	1	2	3	4	5	6
Sequence	2	8	14	20	26	32

Hopefully you had the same numbers for terms 3, 4 and 5 as the ones in my table because you did:

6 x 4 = 24 then take away 4 = 20

6 x 5 = 30 then take away 4 = 26

6 x 6 = 36 then take away 4 = 32

And here's another thing; have a look at the table on the next page and see what I have added.

6n – 4

Term (n)	1	2	3	4	5	6
Sequence	2	8	14	20	26	32

Difference   +6   +6   +6   +6   +6

I can *always* check if I have done the job right, because if I have, then as the rule was **6n** – 4, the sequence should always go up by 6 each time.

By writing the difference between each number in the sequence I can see that the numbers do, in fact, go up by 6 each time.

So, once I have done a question like this (even as a maths teacher) I go back and do my **ABC**.

I don't take my answer for granted. I am **always checking**.

Ok. Draw a blank table and then complete the sequence described by the *nth* term rule below and once you have done that, turn to the next page to check your answer:

4n + 2

4n + 2

Term (n)	1	2	3	4	5	6
Sequence	6	10	14	18	22	26

And a quick reminder of how I got the sequence:

4 x 1 = 4 then add 2 = 6

4 x 2 = 8 then add 2 = 10

4 x 3 = 12 then add 2 = 14

4 x 4 = 16 then add 2 = 18

4 x 5 = 20 then add 2 = 22

4 x 6 = 24 then add 2 = 26

There, that's all there is to it.

I hope by now that you are feeling much more confident about algebra, whether it's solving, expanding, simplifying or the dreaded nth term.

I have tried to go through things in a very detailed, step-by-step way in this chapter so you can actually see, in detail, how the algebra questions on an exam paper work

If you have been able to follow this chapter and work through the examples without too much trouble, then you really are well on your way to nailing the **Up To The Staples** bit of the exam.

If there is anything in this chapter that you are still unsure about or not confident about, then it might be an idea to slowly go back over the chapter again. This time, just do one topic per day instead of the whole chapter in one go. Otherwise, if you are feeling confident, it's time for a tutorial!

Tutorial

The idea was to try to make algebra as simple as possible. I know that so many pupils worry about algebra and I honestly think they, and maybe you, worry needlessly.

I do think that we sometimes make algebra out to be a bit of a 'monster' when it is really quite straightforward.

So, throughout this booklet I have tried to break algebra down into small, simple bits and give you plenty of clear, step-by-step examples of how the algebra you are likely to get, on the exam paper, actually works.

By doing this I hope that you now feel algebra isn't so bad after all. So, a reminder then of the key points:

- When letters and numbers written next to each other it means they are being multiplied. So, something like 5a means 5 x a.

- If we have to work out the value of something, such as 3b + 2a, when a = 4 and b = 2, all we do is put the numbers in place of the letters.

- So, we would work out **3 x 2** + **2 x 4,** which is **6 + 8 = 14**

- Whenever we have to simplify, just think of the T-shirts in the wardrobe and focus on the fact that we are trying to collect things together that are the same.

- So, if the question asks you to simplify **6b + 3a − 4b + 2a**, all you have to do is put things together first of all:

- **6b − 4b + 3a + 2a**

- Then you work out how many there are of each thing:

- 6b − 4b = **2b**

- 3a + 2a = **5a**

- Now put the two answers together:

- **2b + 5a**

- When you have to multiply out, remember to multiply **both** things inside the brackets, so **5(2a + 3)** is **10a + 15**, and **not** 10a + 3

- Sequences are just a matter of checking to see how much is being added or taken away each time, so for a sequence like this:

- 3, 8, 13, 18, 23 we would find out that to get to the next number each time we are adding 5.

- That then tells us what the next 2 numbers will be:

- 23 + 5 = 28

- 28 + 5 = 33

And remember, the main risk with this type of question is from making a daft error, so that is why the **ABC** is so important. Once you have written the answer, go back and check it a second time.

And for a question that uses *nth* term, just use a table like the one we used a few pages ago.

Right then.

Time for a bit of practice!

Some For You To Try

As you work through this page, if you are at all unsure or have any doubts, look back at the examples in this chapter.

1. Solve $5b = 40$

2. Simplify $6b + 3c + 2b + 4c - 3b$

3. Solve $6a = 36$

4. Work out the value of $4h + 3k$ when $h = 5$ and $k = 4$

5. Multiply out $4(2a + 3)$

6. Write the next two terms in this sequence: 7, 11, 15, 19, 23 and explain how you worked out your answer

7. The expression for the ***nth*** term of a sequence is $5n + 3$. Write down the first three terms of this sequence (remember: **use the table**)

8. Solve $4b + 3 = 11$

9. Expand $6(3b - 2)$ **

10.   Simplify $4b + 3c + 5d + 2b + 2c - 2d$ **

The last two questions, marked ** are slightly more challenging than the examples in the chapter, but only *slightly* more challenging!

Some For You To Try - Solutions

1. Solve 5b = 40   Solution: 40 ÷ 5 = 8 so b = 8

2. Simplify 6b + 3c + 2b + 4c – 3b Solution: write things next to each other that are the same: 6b + 2b – 3b = 5b; 3c + 4c = 7c now you can put the two answers together: 5b + 7c

3. Solve 6a = 36   Solution: 36 ÷ 6 = 6 so a = 6 (36 is a square number and if you know that fact, then you would also just 'know' that the answer is 6)

4. Work out the value of 4h + 3k when h = 5 and k = 4

Solution: 4h means 4 x h and 3k means 3 x k so your calculation would be: 4 x 5 = 20 plus 3 x 4 = 12 and your final solution would be: 4h + 3k = 32

5. Multiply out 4(2a + 3) Solution: remember that you have to multiply everything that is inside the brackets (the 2a + 3 bit) by the 4 on the outside of the brackets. So, 4 x 2a = 8a and   4 x 3 = 12 which means that if you put those two answers together, your final answer is 8a + 12

6. Write the next two terms in this sequence: 7, 11, 15, 19, 23 and explain how you worked out your answer.

Solution: the first step is to find out how much is being added to each number to get the next number: As the sequence is going up by 4 each time, the next two terms (numbers) in the sequence are 23 + 4 = 27 and 27 + 4 = 31. Your explanation for how you worked out the answer should be "I added 4" or simply write +4

7 The expression for the *nth* term of a sequence is 5n + 3. Write down the first three terms of this sequence (remember: use the table)

Solution: Start off by drawing a table of values and put the numbers 1, 2 and 3 on the top row. These are the *n* part.

n	1	2	3
5n	5	10	15
+3	8	13	18

So, the first 3 terms (numbers) in the sequence 5n + 3 are the bottom row numbers: 8, 13, 18

8 Solve 4b + 3 = 11

The solution is to work backwards. We know that *something* plus 3 = 11 so start off by taking 3 away. 11 – 3 = 8 which means that the 4b part = 8 and 8 ÷ 4 = 2 and that gives you the answer: b = 2

9 Expand 6(3b – 2) ** This was a slightly more challenging question than question 5 (which was very similar) for two reasons: firstly, I put a minus sign in and secondly I used a different word; Expand. The word expand means exactly the same as 'multiply out' so just do the same as in Q5.

Multiply everything inside the brackets (3b – 2) by the 6 on the outside.

6 x 3b = 18b and 6 x -2 = -12

Put the two answers together and you have 18b - 12

One of the common errors pupils make in a question like this is to write + 12 instead of – 12 because they forget that in the question there was a minus symbol. So, watch out and take care with the signs in your answers.

10 Simplify 4b + 3c + 5d + 2b + 2c − 2d ** The reason
    this was a   more *challenging* question is simply
    because there three letters involved. Otherwise it
    is exactly like Q2:

The solution is still to write things that are the
same next to each other: 4b + 2b + 3c + 2c + 5d −
2d and that gives you a final answer of:
6b + 5c + 3d

Be careful with the last part. Often pupils 'miss'
the minus sign and wrongly write 7d because they
add 5d and 2d instead of subtracting

Some More For You To Try

1. $6b = 42$
2. $9a = 90$
3. $12a = 84$
4. $5b = 12.5$
5. $7b + 3 = 66$
6. $8b - 7 = 9$
7. $2(a + 4) = 20$
8. Simplify $5d + 3g - d - 2g + 7d$
9. Simplify $3a^2 + 4b + 6b + 2a^2 + 5a$
10. Factorise $20d + 12$
11. Factorise $14b^2 + 6b$
12. Multiply out $5(f + 3)$
13. Multiply out $3(k - 5)$
14. Multiply out $2b(b + 7)$
15. Write down the next term in this sequence 8,11,14,17,20
16. Write down the next term in this sequence 2, 8, 14, 20, 26
17. The nth term for a sequence is $6n + 3$. Write down the first three terms in the sequence
18. The nth term for a sequence is $3n - 4$. Write down the first two terms in the sequence

Some More For You To Try - Answers

1. a = 7
2. b = 10
3. a = 7
4. b = 2.5
5. b = 9
6. b = 2
7. 2a + 8 = 20 so a = 6
8. 11d + g
9. $5a^2 + 5a + 10b$
10. 4(5d + 3)
11. 2b(7b + 3)
12. 5f + 15
13. 3k − 15
14. $2b^2 + 14b$
15. 23 (adding 3 each time)
16. 32 (adding 6 each time)
17. 9, 15, 21
18. -1, 2, 5

That's it.

You now have all the basics of algebra that you need to help you achieve your target grade or even a grade C.

If you have reached this point and thought:

"Hey; algebra isn't bad. It's quite easy"

Then I want to tell you what you have to do next:

Practice, practice, practice.

If you become good at a skill and then leave that skill alone, it becomes rusty.

To keep a skill sharp and in good working order you need to use it regularly, so now you must practice, because:

Practice makes permanent.

And if you want any additional resources and materials to help you prepare for your GCSE maths exam, or information about my other GCSE maths books, just visit my website:

www.mathsblasters.co.uk

Best wishes

Graham

The GCSE MathsBlaster